George Kuwayama

February 5 - April 26, 1987

SHIPPŌ

The Art of Enameling in Japan

Far Eastern Art Council of the Los Angeles County Museum of Art

NO. 4

NO. 3

2

NO. 12

NO. 9

Published by the Far Eastern Art Council
of the Los Angeles County Museum of Art
5905 Wilshire Boulevard
Los Angeles, California 90036

Copyright © 1987 by Far Eastern Art Council,
Los Angeles County Museum of Art.
All rights reserved.

Edited by Susan L. Caroselli
Designed by Lilli Cristin
Text set in Kennerly by Continental Typographics
with display type set in Berling by Aldus Type Studio
Printed on Quintessence Dull by Typecraft, Inc.
Cover: No. 1

LIBRARY OF CONGRESS CATALOGING-IN-PUBLICATION DATA
Kuwayama, George.
   Shippō : the art of enameling in Japan.
   Catalog of an exhibition.
   1. Enamel and enameling—Japan—Exhibitions.
I. Los Angeles County Museum of Art. Far Eastern Art Council. II. Title.
NK5004.J3K89 1987 738.4'0952'074019494 86-31839
ISBN 0-87587-136-4

Contents

8 Acknowledgments

12 *Shippō*: The Art of Enameling in Japan

32 Checklist

56 Glossary

NO. 16

NO. 19

NO. 27

NO. 26

NO. 40

NO. 38

Acknowledgments

SHIPPŌ: THE ART OF ENAMELING IN JAPAN is the first historically comprehensive exhibition of this medium in the West. It presents enameled architectural ornaments, sword fittings, and appliqués from the seventeenth and early eighteenth centuries, when technical mastery was combined with an inspired utilization of the inherent aesthetic features of both enamel and metal. Also included are cloisonné vases and panels from the late nineteenth and early twentieth centuries, admired for their virtuosic wire work, opulent colors, and subtle tonalities.

This exhibition is the fortunate recipient of generous loans from many distinguished private collections. I wish to express my profound gratitude to Nobuo Ogasawara of the Tokyo National Museum and to Kazuhiko Inada of the Kyoto National Museum for their kind intercession on our behalf, which enabled us to secure the loans from Japanese collectors. I am deeply grateful to Yuko Hosomi, Kunio Shimano, Takeshi Wakayama, Takashi Yanagi, and the Tokyo National Museum for enamels of the Edo period, and to Karen Chang, Marie Wyatt and Eugene Franke, Mr. and Mrs. Jerry Freeman, Donald Gerber, Dr. and Mrs. Shigeji Takeda, Dr. and Mrs. George Walker, the Philadelphia Museum of Art, and several anonymous lenders for their handsome works of art of the Meiji period. A research grant from the National Endowment for the Arts enabled me to study collections around the world in preparation for this exhibition. Information about private collections was kindly provided by Kay Whitcomb, and reference materials on the cloisonné masters of the Meiji period were lent by Eugene Tosk.

This exhibition and its catalogue were made possible by the magnanimous support of the Far Eastern Art Council of the Los Angeles County Museum of Art. Their encouragement and liberal funding have made this exhibition a reality. Earl A. Powell III, director of the Los Angeles County Museum of Art, provided administrative encouragement for this project. The loan and shipment of the works of art were arranged by John Passi, exhibitions coordinator, and Renee Montgomery, registrar. The catalogue was ably edited by Susan L. Caroselli and designed by Lilli Cristin; the printing was supported by the generosity of Typecraft, Inc. The enamels were photographed by Jeffrey Conley. Additional photographs of objects in Japanese collections were obtained through the good offices of Satoru Sakakibara of the Suntory Art Museum. Photographs of the Hirata school sword fittings were provided by the Tokyo National Museum.

I wish to thank Bernard Kester for the exhibition design and James Peoples and the members of his staff—carpenters, painters, electricians, and museum technicians—for their sympathetic cooperation in the complex tasks of construction and installation. Information on the exhibition was disseminated by the Press Office, headed by Pamela Jenkinson Leavitt. Finally, I would like to convey my deepest appreciation to my secretary-assistant, Karen Nagamoto, for her dedicated efforts.

GEORGE KUWAYAMA
Senior Curator, Far Eastern Art

NO. 47

NO. 53   NO. 52

Shippō: The Art of Enameling in Japan

In Japan the art of enameling has a history spanning many centuries. The sparkle and sheen, brilliant color, and durability of enameled surfaces have endeared them to generations of Japanese patrons. Although the earliest example of enameling in Japan may be dated to the seventh century, it was not until the seventeenth and early eighteenth centuries that the art of enameling matured and attained a peak of aesthetic perfection. In that period consummate technical mastery was combined with the artistic use of both enamel and its metal support, effectively exploiting the inherent qualities of each in the design. Subsequently, in the nineteenth century, new methods introduced from Europe stimulated a revival of enameling with a more brilliant and varied palette of colors. This, combined with refinements in wire techniques, culminated toward the end of the century in superlative cloisonné work admired for its technical virtuosity and artistic achievement.

The Japanese term for enameling is *shippō,* literally meaning "seven treasures," which suggests that enamels may have been created as substitutes for inlaid precious gems. In the process of enameling, a silicate mixture is fused by firing to a metal base, producing a sparkling, colorful surface.[1] Early Japanese enameled works were mainly produced by the champlevé technique, in which enamel is poured into compartments hammered into (repoussé), gouged out of, or cast in metal. The use of wires (cloisons) to form compartments is equally ancient, but cloisonné did not become the dominant method of enameling in Japan until the late nineteenth century, when the precision and delicacy of the pieces were unsurpassed.

Historical Background

ORIGINS OF ENAMELING

The technique of enameling was introduced to the Far East from western Asia and Europe. It developed from the allied art of glassmaking, which may possibly be traced back to the fifth millennium B.C., or earlier, in Egypt and Mesopotamia. Early "proto-enamels" were made of glass inlaid in or fused onto a metal base.

True enameling began in the ancient Near East or in Mycenae. Several pieces of Mycenaean gold jewelry dated to the thirteenth century B.C. contain what have been identified as the earliest enamels, applied to shallow depressions encircled by granulated bands soldered in place. The art continued to thrive in ancient Greece and Scythia and subsequently in Western Europe, Sassanian Persia, Byzantium, and the Islamic world. Extant examples of Sassanian cloisonné attest to a flourishing art of enameling in Iran and south Russia from the third to the fifth centuries,[2] and cloisonné enamels were produced in Constantinople from the fifth century onward. These works were widely disseminated in Eurasia and the techniques of enameling introduced to such disparate lands as Gandhara,[3] Korea, and probably China.[4]

KOREAN PROTOTYPES

The earliest enamels produced in East Asia were discovered in Korea in tombs of the Three Kingdoms period (first century B.C.–A.D. 668). The sumptuous gold jewelry and elaborate tomb furniture found in fifth- and sixth-century tombs indicate a close cultural kinship with the Eurasian steppe, probably the conduit of Western enameling techniques. An enamel-decorated plaque of the late Koguryo period (37 B.C.–A.D. 668) was found at the Chong am ni temple near Pyongyang.[5] Several enameled pendants attached to crowns, earrings, and girdles were excavated from tombs of the early Silla period (57 B.C.–A.D. 668). Gold earrings dated to the late fifth or sixth century had leaf-shaped pendants filled with transparent enamel: the Gold Crown tomb, Kyongju, yielded earrings set with deep blue enamels; the Michu tomb, with green.[6]

These Korean pieces seem to have been made using the technique of subjecting a glass rod to a flame and directing the molten drops into heated metal compartments, a method differing from the more conventional enameling process of adding powdered silicates that fuse when fired. This is suggested by the uneven, convex surface of the enamel.[7]

Early Japanese Enamels

THE KEGOSHI PLAQUES

The earliest extant Japanese enameled works were excavated from the Kegoshi tomb at Sakaimura, Nara, which can be dated by its archaeological context to the latter half of the seventh century.[8] The enamels are found on a group of small, hexagonal bronze plaques once attached to the sumptuous lacquered casket of a Japanese nobleman (FIG 1). Each plaque is adorned with a central rosette of six petals outlined either by ridges cast in the

NO. 60

NO. 66

15 NO. 61

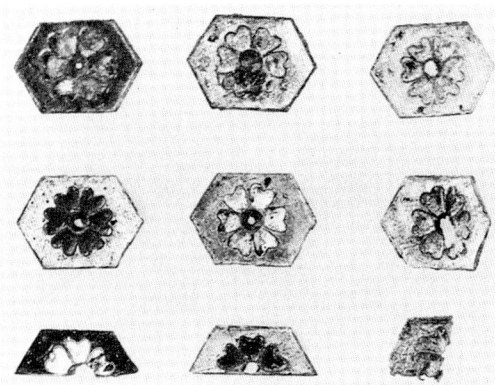

FIG 1 Casket plaques from the Kegoshi tomb mound, second half seventh century. Asuka-mura Educational Commission, Nara. Photo: Jirō Sugiyama, *Toyo kodai garasu*, 1980.

bronze or by applied bronze cloisons, gilded on their exposed edges and filled with translucent golden brown enamel. Some of the plaques have golden brown rosettes in a field of opaque creamy white enamel, others have the opposite color scheme, and still others are embellished with green, brown, or yellow enamels.[9] They were originally affixed to the coffin by metal pegs through a central perforation.

It is difficult to identify the origin of the Kegoshi plaques, an isolated example of seventh-century Japanese enameling. Objects made in the allied medium of glass have been excavated from archaeological sites as early as the Yayoi period (200 B.C.–A.D. 250), and there are many glass decorations on early Japanese Buddhist sculptures of the late sixth and seventh centuries, thus lending credence to the possibility of a native origin for the plaques.[10] The seventh century saw a flourishing cultural interchange between the Korean kingdoms of Paekche and Silla and Japan and close ties between the reigning Japanese aristocracy and the Koreans. The style of the Kegoshi plaques suggests probable Korean inspiration but Japanese manufacture. The convex enamel surface, the irregular protrusions, and the incomplete filling of the compartments indicate that the plaques were made by dropping molten glass into heated metal compartments. This is the same technique as that used for the Korean pieces, which further suggests that they are the technological prototypes for the Japanese enamels.

THE SHŌSŌIN MIRROR

Another early example of enameling in Japan is the mirror preserved in the Shōsōin Treasury of the Tōdaiji,* whose objects are traditionally dated prior to 756 (FIG 2). The Shōsōin mirror was probably made in the official Casting Bureau, which was responsible for creating objects of gold, silver, bronze, iron, lacquer, glass, and beads. The stylistic, iconographic, and technical similarities of the mirror to Chinese, Korean, and Japanese works in metal, lacquer, ceramics, and textiles are consistent with an international style originating in China during the Tang dynasty (618–907) and current during the seventh and eighth centuries throughout East Asia.

The back of the silver mirror is embellished with opaque brown and translucent dark and light green enamels forming a lotus blossom delineated with gold and gilt silver cloisons.

*The suffixes "-ji" and "-in" denote temples.

The petals of the lotus were formed by cutting silver sheets into shallow pans whose raised edges formed cloisons; inner contour lines were created with additional silver wires. Molten glass of the desired color was then poured into the pans. The enamels were left unpolished in concave pools between the wires (the concavity is indicative of close adhesion to the metal walls, confirming a high lead content in the enamel). Eighteen petal forms and twelve gold bordering sections were then affixed to the mirror with lacquer.[11]

THE BYŌDŌIN DOOR FITTINGS

Various literary sources of the Heian period (794–1185) refer to objects adorned with *shippō*, but, because of the ambiguity of the language, it is far from certain that all of these were actually enameled works.[12] One example involves the door fittings of the Hōōdo, the main hall at the Byōdōin, dedicated in 1052. These iron door latches were adorned with floral rosettes consisting of four petals surrounding a central stamen; the four-petal design is found elsewhere in the Hōōdo and is typical of the Byōdōin. According to Motoo Yoshimura, this design occurred in two varieties, one with petals in blue-green enamel and the other in copper inlay. He noted the similarity of materials between the enameled door fittings and glasslike decorations on the Hōōdo canopy, suggesting that the technique utilized here was dropping molten glass onto metal as in the Shōsōin mirror and the Kegoshi plaques.[13]

A dissenting observation was subsequently made by Norio Suzuki, who reported that the Byōdōin door latches were ornamented only with copper inlay. He went on to claim that green copper corrosion had mistakenly been identified as enamel.[14] In yet another view, Mitsuhito Mori suggested that the Byōdōin door fittings employed crushed glass as the basic ingredient for enamel and that these fittings comprise the earliest Japanese examples of champlevé iron.[15] Perhaps this dilemma can be resolved, in the absence of any technical analysis, by the suggestion that some of the fittings are original works of the eleventh century while others may be restorations or replacements from a later period and are in fact executed in a variety of techniques.

MEDIEVAL ENAMELS

After the Heian period there is no mention of enameling in textual sources until the

FIG 2 Mirror back, first half eighth century. Shōsōin Treasury, Tōdaiji, Nara. Photo: Norio Suzuki, *Nihon no shippō*, 1979.

Higashiyama period in the second half of the fifteenth century, when Chinese cloisonné vases were imported for Shōgun Ashikaga Yoshimasa's enjoyment.[16] If enameled works were made in Japan during the medieval period, they were probably created by glass craftsmen using the molten glass technique. Chinese objects with enameled decorations were made in a different manner: the artisans placed powdered silicates in areas defined by wire cloisons, heating and finally polishing the finished enamel. This was the way cloisonné enamels were manufactured in Europe and the Near East, a technology that had been introduced into China during the Yuan dynasty (1271–1368)[17].

This cloisonné technique was adopted by the Japanese by the Momoyama period (1573–1615), but the exact source of the new process in Japan is unknown. A Korean origin has been hypothesized, part of the aftermath of Shōgun Toyotomi Hideyoshi's abortive conquest of Asia in the 1590s.[18] However, few, if any, enameled works can be attributed to sixteenth-century Korea with any certainty. A far more probable source, although as yet unsubstantiated, was contact with China itself either directly or through Portuguese intermediaries during the second half of the sixteenth century.

The renowned tea master Sōami (1485–1525) used cloisonné extensively in his furnishing of the Higashiyama palace reception hall. The *Kundaikan sayu chōki* (A fifteenth-century connoisseur's manual of Chinese art) records that the hibachi coal tongs and wine cups and stands were all in cloisonné; it even illustrates the cups and stands.[19] Evidently, enameled wares were much prized during the Muromachi period (1333–1573): the *Onryōken nichiroku* (Daily record of the Onryōken*) relates that Ashikaga Yoshimasa visited the Shōsenken of the Shokokuji, Kyoto, in 1462, after which he sent a gift of a pair of *shippō ruri* (lapis blue–enameled) flower vases.[20]

Seventeenth-century Enameling

MOMOYAMA ENAMELING

It was not until the beginning of the seventeenth century, during the Momoyama and Edo (1615–1868) periods, that enameled works were made in substantial numbers. Utilizing the new technique of applying powdered enamel to a cold metal ground, artisans produced fittings for swords and chests, writing accoutrements for scholars' desks, and architectural embellishments such as nail covers and door pulls for sliding screens.

*The suffix "-ken" denotes a hall.

FIG 3 Crucifix, Momoyama period (1573–1615). Osaka Municipal Art Museum. Photo: Motoo Yoshimura, *Shippō*, 1966.

The Momoyama spirit of grandeur that produced magnificent castles and colorful paintings on glittering gold screens provided an ideal milieu for brilliantly enameled fittings on buildings and furniture. An exuberant taste for gorgeous color and fine metalwork led to an era of remarkable enameling that continued through the Genroku period in the early eighteenth century.

Two nail covers (NOS 1 and 3) and a cover for a hand warmer (NO 2) in the exhibition have been attributed to the Momoyama period.[21] One, decorated with the Toyotomi crest of paulownia delineated with cloisons, is considered to have come from the Jurakudai, Hideyoshi's palace, built in 1587. The famous tea master and connoisseur Kobori Enshū (1579–1647) is said to have produced a suit of armor embellished with enameled designs. A Christian crucifix (FIG 3) adorned with polished white, brownish red, green, and purplish blue enamels on a copper ground may also be ascribed to the Momoyama period. A terminus date of 1614 may be ventured for the creation of Christian icons, which were proscribed in Japan after that year. The techniques used to make the crucifix seem similar to those of China in the Wan-li period (1573–1619), which has prompted speculation on a possible Chinese manufacture.[22]

EARLY EDO ENAMELING

Construction of the Tōshōgū shrine in Nikkō, a mausoleum for Shōgun Tokugawa Ieyasu (1542–1616), began in 1616 and continued for three years. Major additions were initiated in 1634; a number of extant metal fittings decorated with enameling, especially opaque ruby red and green, date from this period. The construction records, *Tōshōgū gosoeichō*, mention fittings with cloisonné in a variety of techniques and colors. They also name the metalsmiths Echizen, Sandaifu, and Sonjuro.[23]

In 1634 the shōgun Tokugawa Iemitsu (1603–1651) visited Nagoya Castle, and its *jorakuden*, the residential hall, was rebuilt for his use. The palatial quarters constructed in the classic *shoin zukuri* architectural style had sliding screens with paintings by Kanō Tan'yu (1602–1674). The door pulls on these screens were made of red copper and decorated with a gilded crest of the Tokugawa family and chrysanthemum border patterns in white and green enamel.[24] Nail covers were also made utilizing similar techniques. While

FIG 4 *Tsuba* by Hirata Hikoshirō, called Dōnin (1591–1646), early seventeenth century. Important Cultural Property, collection of Tadanao Yoshii. Photo: Norio Suzuki, *Nihon no shippō*, 1979.

there is speculation about the origin of some of the earlier Momoyama pieces, these are unquestionably of Japanese manufacture.

HIRATA DŌNIN

Hirata Hikoshirō, called Dōnin (1591–1646), traditionally accepted as the creator of the Nagoya Castle *jorakuden* door pulls and the Tōshōgū nail covers, is said to be the first Japanese master to work with true enamel rather than molten glass.[25] Japanese literary sources credit him and his numerous descendants for the introduction and the development of the modern art of enameling. The *Kotō kinkō meipu* (Genealogy of metalsmiths in the capital), published in 1810, states that a Korean artisan taught Dōnin how to apply enamels to metal during the Keicho era (1596–1614),[26] but there is little evidence of an advanced enameling industry in Korea at this time. It is possible that Dōnin's Korean mentor learned from the Chinese.

Dōnin was appointed maker of sword fittings and *shippō* for the Tokugawa shōgun. Although many enameled works are only attributed to him, there is one work that is widely accepted to have been made by his hand: an exquisite sword guard (*tsuba*) in the collection of Tadanao Yoshii (FIG 4). Silver cloisons define clouds and floral patterns with great delicacy, and green, red, black, light celadon green, and opaque white enamels fill the compartments.[27]

The Hirata clan became hereditary metalsmiths and enamelers to successive shōgun, concentrating their talents on sword fittings (NOS 25–29). They crafted amazingly beautiful *tsuba* and *mitokoromono* (fittings) with elegant designs and refined detailing, and produced, with a stunning technique, complex cloud and floral patterns in a varied palette of red, green, blue, black, and white enamels. Few signatures appear on works by the first four generations of Hirata artists; however, the fifth, Narikado, often signed his works. After Hirata Harunari (d. 1840) (NOS 31–32), the eighth-generation descendant of the clan, the main Hirata line disappeared and collateral Hirata families continued a declining tradition. In the cities, metalsmith-enamelers outside the official Hirata clan produced works for prosperous merchant clients.[28]

KACHŌ

Another major figure in the development of enameling in Japan was Kachō, from Matsuyama, Shikoku. He was invited to Kyoto by Hideyoshi and worked there at the same time as Dōnin. The door pulls of Katsura Palace are traditionally ascribed to Kachō. The palace is attributed by a popular legend to Kobori Enshū; whether he actually directed the construction is uncertain, but its design was at least influenced by his style. Enamels were often used in buildings constructed under his supervision; this preference is confirmed by his personal inventory, *Enshū kōokuracho,* which mentions enameled objects.[29] According to one account, Kobori often requested Kachō to create enamels for him.

There are three periods of construction at Katsura, beginning in 1620, 1641, and 1656. The Ninomado in the Shōkintei (pavilion) of Katsura was constructed in the early 1640s,[30] and the sliding screens painted by Kanō Tan'yu have cloisonné door pulls made by Kachō with a unique coiled shell pattern. Champlevé techniques are employed in the enamel door pulls of the later structures at Katsura, such as the Jōdannoma (reception room) of the Shin goten.

SEVENTEENTH-CENTURY CHAMPLEVÉ ENAMELS

Enameled works created during this period were primarily in champlevé with insets gouged with chisels or hammered out in repoussé; the use of cloisons was rare. The enamels were usually opaque and lacked the sheen characteristic of later pieces. A number of dated objects remain from the mid-seventeenth century, providing a corpus of stylistic monuments. A striking example is a lacquered wood saddle with gilt bronze fittings decorated with floral patterns filled with green and red champlevé enamels; a date corresponding to February 1636 is inscribed in the wood on the underside of the saddle.[31]

The Kongōbuji ancestral shrine at Koyasan, completed in 1643 and dedicated to Tokugawa Ieyasu and Tokugawa Hidetada (1579–1632), displays turquoise champlevé enamels decorating nail covers and metal fittings.[32] The Ryūkoin of the Daitokuji, completed in 1644, has nail covers with champlevé designs filled with celadon green and brownish red enamels. In one of the rooms, constructed in 1641–42, there are cupboard

FIG 5 Document chest lock, mid-seventeenth century. Tōshōgū, Nikkō. Photo: Norio Suzuki, *Nihon no shippō,* 1979.

doors, painted by Shōkadō (1584–1639), whose door pulls are embellished with champlevé enamels. These enamels may be dated to this period and associated with Kobori Enshū and Dōnin, who were both still active during these years.[33]

A document chest made of gold *maki-e* lacquer, in the Tōshōgū shrine at Nikkō, has gilt silver fittings embellished with translucent champlevé enamels of celadon green, blue, brownish red, and white (FIG 5). The Tokugawa emblem on the chest has forty-two stamens, a design feature datable to before 1650. Hirata Narihisa (d. 1671) is thought to be the artist who made these fittings, probably in the late 1640s, since document chests came into fashion after 1646.[34]

A matched pair of lanterns in the Daisonin hall of Nikkō carries the inscription of the metal craftsman Horiyamajō Kiyomitsu and the date 1653. These memorial lanterns, dedicated to Tokugawa Iemitsu, were lavishly embellished with opaque polychrome champlevé enamels in a rich palette of hues including white, green, yellow, black, blue, and reddish purple. One of the pieces preserved in the Tōshōgū shrine's Okumiya Sakashita Mon (gateway) has enamels of two different colors placed side by side without cloisons yet producing a precise pattern.[35]

In 1656 at the Manjuin in Kyoto twenty-five nail covers were made in two varieties, one in the form of Mount Fuji, with black enamels applied to a champlevé carved copper ground, and the other cloud-shaped with green and turquoise enamels. In contrast to the ostentatious brilliance of Nikkō Tōshōgū enameling exemplified by the previously mentioned pieces, the Kyoto style is simple and the colors more subdued.[36] Similar in style to the Manjuin examples are the superb nail covers at the Kuroshoin of the Nishi Honganji, datable to 1657. Blue, red, and green enamels were applied to metals handsomely shaped by champlevé and repoussé techniques.[37]

During the Edo period enameling was an art executed principally by metalsmiths, which accounts for the artisans' great sensitivity to metal forms. Colorful enamels were selectively applied to areas where they would function most effectively in the overall design. This harmonious relationship between enameling and metalwork became an established

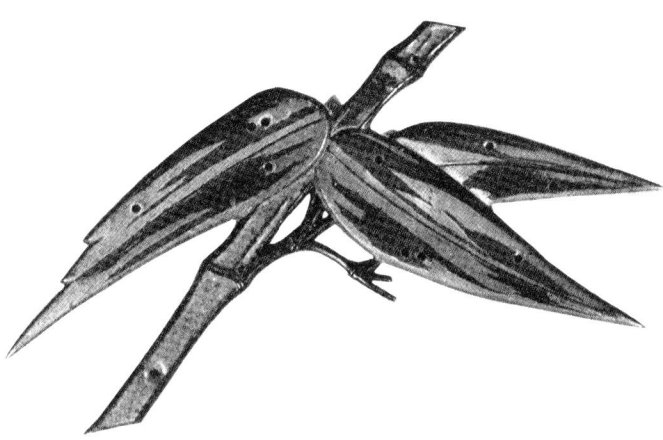

FIG 6 Appliqué, 1659. Shugakuin, Kyoto. Photo: Motoo Yoshimura, *Shippō*, 1966.

feature of the mid-seventeenth-century style. The nail covers of the Kuroshoin as well as those at the Shugakuin palace, built in 1659, epitomize this refined artistry.[38] At the Shugakuin there is an appliqué of bamboo leaves in brown enamels highlighted with golden streaks attached to a copper stalk decorated with blue enamels (FIG 6), and a flower cart on the nail covers of the Kyakuden (chambers) is rendered in skillfully worked copper repoussé, with finely delineated red, blue, and white champlevé enamel camellias and peonies.

Eighteenth-century Enameling

The period from the mid-seventeenth century through the early eighteenth century was a halcyon era for the artistic creation of enameled objects. Its apogee was in the early eighteenth century, a time that encompassed the complete technical mastery of enameling and the disciplined use of the medium toward artistic ends. The *Soken kisho* (Appreciations of superior sword furnishings) of 1786 lauds as the greatest master of *shippō* Hirata Narikado (1684–1757), whose works were superior to Chinese imports and among the best produced by the Hirata clan. His large atelier trained an extraordinarily talented group of apprentices.[39]

One of the finest creations in early eighteenth-century Japan was the Onari goten in Tokyo, an extremely elaborate palace complex of forty-eight buildings constructed in 1702 for Lord Maeda Shōunkō. Kazariya Seizaemon, one of its principal designers, had a preference for metalwork and may himself have designed its enamel nail covers (FIG 7). A notation in the *Sangikō nenpyo* (Councillors' chronological records) lists his name with those of the artisans Matsui Yashichirō, Hachiami Genshirō, and Tanami Genshirō, who may all have participated in the creation of these works.[40] They exemplify the high standards of workmanship of the period in both enamel and metal. Flower baskets or bird and cricket cages of repoussé and engraved gilt copper were set with opaque blue, red, green, and purple champlevé enamels in designs that matched adjoining paintings of birds and flowers by Kanō School masters.

During the late eighteenth century enameled works were made for a broader range of patrons and displayed rich combinations of color. Although fewer enamels seem to have been produced, two fine datable works from the later half of the eighteenth century may be cited. Enameled images of the Zen patriarchs Kanzan and Jittoku adorning an iron spear have contour lines defined by cloisons containing turquoise, white, black, and reddish brown enamels. In contrast to the opaque enamels usually found on earlier examples, the turquoise areas are translucent. An inscription inside the storage box reads, "A lucky day in March, 1757, Yamagataya, Ichibei." A major historical monument of the late eighteenth century, the Sumiya, the famous teahouse in Kyoto, has bronze door pulls on its sliding screens adorned with enamels of pure white, blue, green, yellow, and black separated by brass cloisons.[41]

Modern Enameling

KAJI TSUNEKICHI

The late Edo years saw a decline in the quality and quantity of enamels. Even the Hirata school, the exclusive purveyors of sword fittings to the daimyō and shōgun, and thus the primary producers of enameled works, suffered from reduced patronage. Fortunately this ancient art was revitalized by the works of Kaji Tsunekichi (1803–1883) toward the end

FIG 7 Nail cover from the Onari goten, Tokyo, c. 1702. Important Cultural Property, collection of Maeda Ikutokukai. Photo: Motoo Yoshimura, *Shippō,* 1966.

of the Tokugawa era. Kaji is traditionally considered the founder of modern Japanese enameling. The second son of an Owari samurai, he became a metal craftsman. He developed his techniques after six years of studying ancient texts and examining "Dutch" enameled ware (actually a piece of Chinese cloisonné that he acquired in 1832), which he fragmented, the better to analyze its properties. His work, not surprisingly, appears to be influenced more by Chinese models than the enameling traditions of Edo Japan.

One of Kaji's important accomplishments was the creation of enameled works in the round, a form far more challenging than two-dimensional plaques. He also developed the use of delicate, ribbonlike silver threads attached to the metal base as cloisons and improvised the means to set cloison wires into enamel flux before firing rather than soldering them to the metal base. In 1838, working independently of the secretive Hirata clan, he succeeded in making his first freestanding cloisonné enameled piece, and in 1850 he began creating works for the Owari daimyō, who subsequently presented handsome cloisonné to the Tokugawa shōgun at Edo. To accommodate his increasing commissions, he accepted apprentices in 1853. Five years later he was producing for foreign buyers.

Early pieces by Kaji have thin-walled copper bodies elaborately decorated with somber-toned enamels; the enamels themselves are speckled with impurities and uncontrolled in their placement within the compartments. Cloisons were added throughout the background to stabilize the enamel, and repetitive patterns were utilized to permit ease of manufacture. With time he perfected his technique and imitated the tightly integrated designs of Japanese textiles, which he applied to heavy bronze bodies. The bases of most Kaji pieces bear a single coat of dark green enamel that was sometimes crudely applied; this color is typical of Japanese enameled wares of the nineteenth century.[42]

The most lasting contribution by Kaji to the art of enameling is his influence as teacher and mentor. The major masters of modern enameling in Japan were all his students or their disciples through several generations. Not only did he create a new artistic movement through his work but trained the artists for a new age. Although Kaji revitalized the moribund art of enameling it was not until the beginning of the Meiji period (1868–1912) that major innovations initiated another era of florescent development for this art. The invention of new colors and refinements in the cloisons eventually led to the production of artistic creations that rivaled the finest porcelains in craftsmanship and subtle beauty.

TSUKAMOTO KAISUKE

Tsukamoto Kaisuke (1828–1897) learned enameling from Hayashi Shōgorō, who had been apprenticed to Kaji Tsunekichi. With Kaisuke a new generation of enamelists arose. The Chinese-influenced style of Kaji was replaced in the mid-1860s by an aesthetic that was more traditionally Japanese. Artisans created shapes that resembled Japanese ceremonial vessels, and they took patterns from native brocades and ceramics and pictorial designs from paintings.[43]

Kaisuke discovered how to fire enamels on large wares successfully and in 1868 was the first to apply cloisonné enamels to a ceramic body. Work with ceramic bodies led to the invention of new, softer colors and more refined and flexible wire work, improvements that were essential for the subsequent development of a naturalistic style.

From the mid-1860s onward it was production for export that provided the economic foundation for a flourishing cloisonné industry in Japan. Kaisuke's student Hayashi Kodenji (1832–1911), who was later to create cloisonné admired for its remarkable naturalism and intricate detail (NO 48), was among the first to sell his work in Yokohama, the port city open to foreign trade.[44]

Kaisuke experimented with pictorial designs in the enamel medium and about 1865 successfully created a scene of Nagoya Castle in the Yamato-e painting style.[45] This work marks the beginning of a new native style of cloisonné, although a "brocade style" derived from textile designs, with dark green backgrounds and red, white, blue, and lilac enamels, was the dominant mode in the late 1860s and 1870s (NOS 44–45). Further development of pictorial scenes naturalistically rendered had to wait for the technical improvements of the late 1870s.

In 1875 the trading firm of Ahrens & Company built a cloisonné factory in Tsukiji, Tokyo, and Kaisuke was invited to work there with Gottfried von Wagner, a German chemist. During the three years of the company's existence Wagner introduced many improvements in the preparation and application of enamels: he replaced dull opaque enamels with sparkling, colorful ones; he introduced new chemicals to brighten the enamels and augment the palette of colors and shades (most notable is a sky blue); and he controlled firing temperatures to prevent body warpage and refired pieces to improve surface luster.

NAMIKAWA YASUYUKI

Perhaps the most celebrated Japanese enameler was Namikawa Yasuyuki (1845–1927) of Kyoto, who left the service of Prince Fushimi for a career as an artist. In the early 1870s he devoted himself to the making of cloisonné. His creations are admired for stunning naturalism, technical excellence, and exploitation of the aesthetic potential of cloisonné, as well as for the refined elegance of his wire work and the subtle beauty of his colors (NOS 49–57). Yasuyuki also introduced important technical innovations: the invention of brilliant mirror-black enamels and transparent enamels in 1879. By 1893 his works were free of supporting background wires, allowing open areas unobstructed by cloisons. This successfully culminated the long process of experimentation that had begun about 1875.

NAMIKAWA SŌSUKE

Yasuyuki's Tokyo contemporary Namikawa Sōsuke (1847–1910)—no relation—was equally famous. He preferred pictorial subjects, and in 1881 he first began reproducing classic Japanese paintings in cloisonné enamel in a remarkably naturalistic style. Sōsuke was especially lauded for his cloison-less enamels *(musen shippō)*, in which he firmly controlled the placement and flow of glazes. By 1889 he and his assistants were creating wireless works of such precision and subtlety of tonal gradation that they were like brush paintings in enamel (NOS 58–60).

THE GOLDEN AGE

The period from 1880 to 1914 was the golden age for modern Japanese enameling. A vast array of technical innovations allowed the artist to exploit his medium fully, utilizing a wide range of colors from delicate pastels to rich, dark hues. Designs ranged from the decorative, with abstract patterns, to highly naturalistic compositions of birds and flowers, realistic renderings of perceptively observed nature. The flourishing market for enamels encouraged artists to indulge in lavish detail, painstaking workmanship, and the use of opulent materials. These enameled works were made primarily for affluent Western patrons, who were attracted by their flamboyant designs and marveled at their flawless technique. Exquisitely made vases, some of which grace this exhibition, were created with great care for discerning patrons.

Technical innovations continued. *Moriage* enamel, rendered in sculptural relief, was developed by artisans Hattori Tadasaburō (active c. 1900), Kawade Shibatarō (1856–1921), and Andō Jūbei (1875–1953) in 1903. Transparent plique-à-jour enamel, resembling stained glass, was imported from France and adapted in Japan before 1910 by Shibatarō and Jūbei in their own unique version.

About 1894 Japanese cloisonné became the craze of Europe; the decade from 1894 to 1905 was the heyday of cloisonné export. Every fashionable home in Europe and America had a Japanese cloisonné dish to receive calling cards. Enormous quantities of enameled works were produced for export at Toshima, Shippōmura, near Nagoya.

The thriving era of cloisonné production ended with World War I. The great artists of the "golden age" were gone, and the creative fervor of the previous decades had ebbed. International interest in cloisonné declined, although enameled wares adopting the styles and traditions established by earlier masters continued to be produced for a commercial market.

Contemporary Enameling   Within the last two decades there has been a renaissance of interest in the art of enameling. The impetus for innovation and creativity has come not only from artists in Tokyo, Nagoya, and Kyoto but throughout Japan. Enormous strides in chemistry have enabled enamelers to produce new permutations of colors and shades, as well as striking blends of hues with suffused tonalities. Technological advances in electric kilns and precision temperature controls have increased the artistic possibilities for the cloisonné artist. The myriad developments in contemporary art in Europe and America have stimulated the enameler to explore new forms and themes, resulting in artistic creations worthy of international attention.

NOTES

[1] Enamel is made from a mixture of silica (sand, flint, or quartz), potash or soda, lead, and borax. The addition of metallic oxides—particularly cobalt, copper, iron, or manganese—imparts a variety of colors. Firing to a temperature of 800° C (1,480° F) fuses the enamel to a base made of copper, silver, gold, or steel.

[2] M. Rostovtzeff, *Iranians and Greeks in South Russia* (New York: Russell and Russell, 1922), 181–82, 190; and Karl Jettmar, *Art of the Steppes* (New York: Crown Publishers, 1964), 61–62. Numerous examples of south Russian polychrome enameled metalwork are extant. Excavated specimens now in the Hermitage, Leningrad, include a gold torque from Kul Oba with bright green and blue enamels, dated to the fourth century B.C., and a shield ornament from Kelermes in the form of a panther with colored glass paste, from the early sixth century B.C. See "From the Lands of the Scythians," *The Metropolitan Museum of Art Bulletin* 32, no. 5 (1975), 101, pl. 5 no. 28, and 113, pl. 20 no. 83.

[3] John Marshall, *Taxila* (Delhi: Motilal Banarsidass, 1951), 134, 606 (427), and pl. 185 (s).

[4] As the traditional center of civilization in the Far East, China is the logical place to search for cultural and technological origins. Glass appears in China as early as the Warring States period (481–222 B.C.). Glass beads have been excavated from Han dynasty tombs (206 B.C.–A.D. 220), and ancient Chinese texts refer to their manufacture during Emperor Wu's reign (140–86 B.C.). The *Beishi* (History of the Northern Wei dynasty) (286–535) mentions the introduction of colored glass by Indo-Scythians about 430. Nevertheless, early examples of Chinese enameling have yet to be discovered. The only extant Chinese works that can be considered are either inlaid with cut glass or filled with powdered glass in a lacquer vehicle. See Jirō Sugiyama, *Toyo kodai garasu* (Ancient oriental glass) (Tokyo: Tokyo National Museum, 1980), 184; Joseph Needham, *Science and Civilisation in China* (Cambridge: Cambridge University Press, 1962), vol. 4, pt. 1, sect. 26, 108; and Sueji Umehara, "Chūgoku shutsudo no hariden (shippō) no ihin" (Glass paste [cloisonné] relics excavated in China), *Museum*, no. 185 (August 1966), 21–24.

[5] Sueji Umehara and Ryōsaku Fujita, *Chōsen kobunka sōkan* (Collectanea of ancient Korean culture) (Tamba: Yotokusha, 1947), vol. 4, pl. 41 no. 115; and Umehara, "Chūgoku shutsudo no hariden (shippō) no ihin," 22 fig. 1 (right).

[6] Archaeology collection, National Museum of Korea, Seoul.

[7] Dorothy Blair, *A History of Glass in Japan* (Corning, New York: Corning Museum of Glass/Tokyo: Kodansha International, 1973), 370–71. Some contemporary enamelers are skeptical about this method, maintaining that the molten glass would spatter and crack unless metal and enamel were at the same temperature. They cite wet packing, a technique in common use today, as the only feasible one for these

pieces. In this process, moistened powdered enamel is packed into the compartment, which is then heated.

[8] Mombushō Bunkachō, *Jūyō bunkazai* (Important cultural properties) (Tokyo: Mainichi Newspapers, 1976), 129, no. 93.

[9] Sugiyama, *Toyo kodai garasu,* 32 fig. 88, 145; and Blair, *History of Glass in Japan,* 343, 370–71. Blair describes the metal outlines of the six petals as cloisons, while Sugiyama refers to them as cast parts of the body.

[10] Early Japanese familiarity with glassmaking is confirmed by the excavation of glass objects at Yayoi sites (200 B.C.–A.D. 250) as well as from those of the Tomb Mound period (250–552). Glass decorations grace Japanese Buddhist sculptures such as the Yumedono Kannon and the Kudara Kannon of the Asuka period (552–645) at the Hōryūji. Ancient texts refer to the numerous glass beads that in 734 adorned the Nishi Kondō of the Kōfukuji. See Sugiyama, *Toyo kodai garasu,* 184–86.

[11] Helmut Brinker and Dietrich Seckel, "The Cloisonné Mirror in the Shōsōin," *Artibus Asiae* 32 (1970): 315–35; and Blair, *History of Glass in Japan,* 342–43.

[12] Blair, *History of Glass in Japan,* n. 140.

[13] Motoo Yoshimura, *Shippō* (Enamelware) (Kyoto: Maria Shobo, 1966), 9.

[14] Norio Suzuki, *Nihon no shippō* (Japanese enamels) (Kyoto: Maria Shobo, 1979), 205.

[15] Mitsuhito Mori, *Shippō bunkashi* (The cultural history of enamelware) (Tokyo: Kondo Shuppansha, 1982), 48–49.

[16] Suzuki, *Nihon no shippō,* 206.

[17] Harry Garner, *Chinese and Japanese Cloisonné Enamels* (London: Faber and Faber, 1962), 31.

[18] Using a technique similar to cloisonné, Korean artisans since the Koryo era (935–1392) have inlaid wires into lacquer chests and boxes.

[19] Yoshimura, *Shippō,* 11.

[20] Blair, *History of Glass in Japan,* 155.

[21] Ibid., 398–99.

[22] Yoshimura, *Shippō,* 12, pl. 65.

[23] Ibid., 13, pls. 4–5.

[24] Ibid., 12, pl. 66.

[25] Dōnin is not to be confused with a contemporary metalsmith in Kyushu also named Hirata, who worked for the Hosokawa domain.

[26] Yoshimura, *Shippō,* 15. The *Sanko furyaku* (Manual of engineering) of 1844 also mentions

the Korean source of Dōnin's enameling techniques; see Suzuki, *Nihon no shippō,* 209.

[27] Yoshimura, *Shippō,* 5 fig. 4. Wide acceptance is given to eight sword fittings—four *tsuba* and four *kozuka*—as the work of Dōnin.

[28] Suzuki, *Nihon no shippō,* 209–10, 255; and James L. Bowes, *Notes on Shippō* (London: Kegan Paul, Trench, Trubner & Company, 1895), 79–109.

[29] Yoshimura, *Shippō,* 16.

[30] Akira Naito, *Katsura, a Princely Retreat* (Tokyo: Kodansha International, 1977), 106–7.

[31] Suzuki, *Nihon no shippō,* 214–15.

[32] Ibid., 215.

[33] Yoshimura, *Shippō,* 17.

[34] Ibid.

[35] Ibid., 18.

[36] Suzuki, *Nihon no shippō,* 215–16.

[37] Yoshimura, *Shippō,* 20.

[38] Ibid., 21, pls. 21–22.

[39] Suzuki, *Nihon no shippō,* 210.

[40] Yoshimura, *Shippō,* pls. 36–38, 95–104; and Suzuki, *Nihon no shippō,* 216.

[41] Suzuki, *Nihon no shippō,* 216–17. The Sumiya enamel door pulls were made sometime between the terminus dates of 1756 and 1838; the style of the door pulls would suggest the late eighteenth century.

[42] James L. Bowes, *Japanese Enamels* (London: Bernard Quaritch, 1886), 41–45, pl. IV.

[43] Ibid., 47–50, pl. V.

[44] There were four generations of Hayashi Kodenji, the fourth of whom died only recently.

[45] Lawrence A. Coben and Dorothy C. Ferster, *Japanese Cloisonné* (Tokyo: Weatherhill, 1982), 59.

Checklist

Momoyama and Edo Enamels

Seventeenth Century

1 NAIL COVER
Momoyama (1573–1615)
Champlevé and cloisonné; gilt copper alloy
5 1/8 x 10 1/2 x 9/16 in. (13 x 26.7 x 1.4 cm)
Private collection, Osaka

The nail cover is constructed in three layers of copper alloy: a flat base sheet, a layer of repoussé, and a layer of gilt metal worked in repoussé and *nanako* (round punch marks made with a cold chisel). The enamels are red, white, dark green, and turquoise.

2 COVER FOR A HAND WARMER
Momoyama (1573–1615)
Champlevé; gilt copper alloy, silver
4 7/8 x 9 13/16 in. (12.4 x 24.9 cm)
Private collection, Osaka

A fish weir amid silver waves and a Hideyoshi crest (*mon*) are juxtaposed. The repoussé metal ground is inlaid with dark red, white, green, and light blue enamels. The crest is represented in dark green enamel and gilt metal stippled with *nanako*.

3 NAIL COVER
Momoyama (1573–1615)
Cast champlevé; gilt copper alloy, silver
4 x 11 in. (10.2 x 28 cm)
Private collection, Osaka

A light blue enameled sheet underlies the lattice pattern of a gilt bronze fish weir. The rolling silver waves are accented with blue, white, red, green, and purple enamels.

1

3

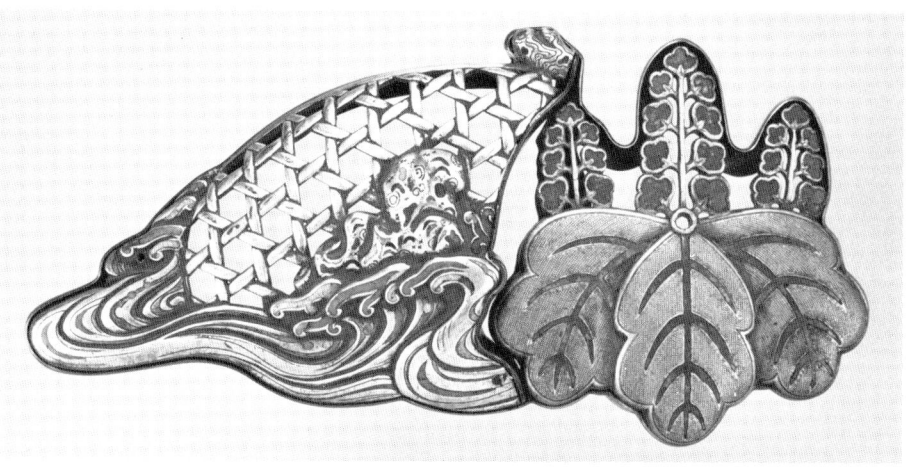

2

4 DOOR PULL
First half seventeenth century
Cloisonné; copper alloy
2 5/16 x 3 7/8 in. (5.9 x 9.9 cm)
Private collection, Osaka

The door pull in the shape of paulownia is made of a cast copper alloy base with light blue, mustard, white, and dark red enamels.

5 DOOR PULL
First half seventeenth century
Champlevé; gilt copper alloy
3 3/8 x 3 3/8 in. (8.6 x 8.6 cm)
Private collection, Osaka

This door pull, assembled from three pieces, is in a tortoiseshell pattern. The outer tier is set with green enamel, and a three-petaled blossom in blue and red enamels forms the center.

6 APPLIQUÉ
First half seventeenth century
Champlevé; gilt copper alloy
1 11/16 x 3 15/16 in. (4.3 x 10 cm)
Private collection, Osaka

White and red enamels represent prunus blossoms, while dark green accents the stems.

7 DRAWER PULL
First half seventeenth century
Cloisonné; copper alloy
1 x 4 15/16 in. (2.6 x 12.6 cm)
Private collection, Osaka

White, green, yellow, and blue enamels are applied to metal in the shape of narcissus flowers; they run together at the ends of the flowers to produce subtle suffusions of color. The smooth surface appears to be polished.

8 NAIL COVER
First half seventeenth century
Champlevé; gilt copper alloy
3 1/4 x 3 5/16 in. (8.3 x 8.4 cm)
Private collection, Osaka

Dark green enamel and fine *nanako* characterize this nail cover with a heraldic design.

5

4

7

6

8

9   WATER DROPPER
First half seventeenth century
Champlevé; gilt copper alloy
3 x 5 x ½ in. (7.6 x 12.7 x 1.7 cm)
Collection of Takashi Yanagi

Green, turquoise, red, and white enamels fill cells created by fine cold chisel work, forming the colorful wing feathers of two cranes. The reservoir of the dropper is textured with tiny *nanako*.

10   WATER DROPPER
First half seventeenth century
Champlevé; gilt copper alloy
3 ¾ x 2 9/16 in. (9.5 x 6.5 cm)
Collection of Takashi Yanagi

Tagasode, a traditional popular theme depicting kimonos, is rendered here by crossed sleeves embellished with red, blue, green, and white enamels. The *shibori* (tie-dyed) pattern is indicated by red and white dots.

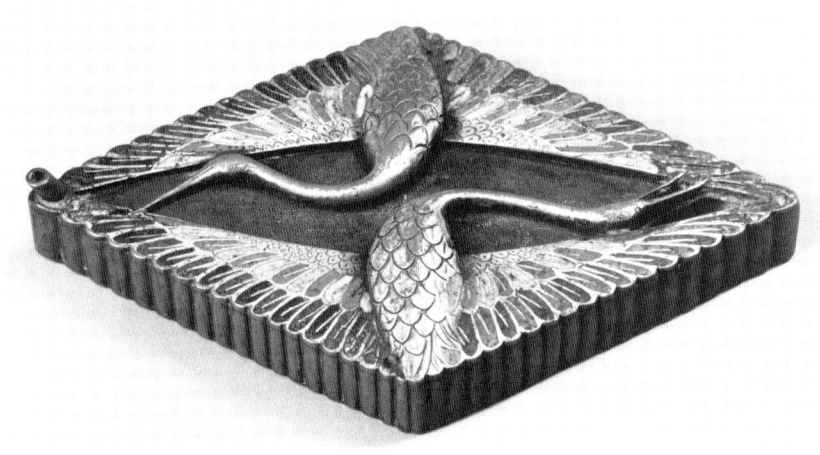

9

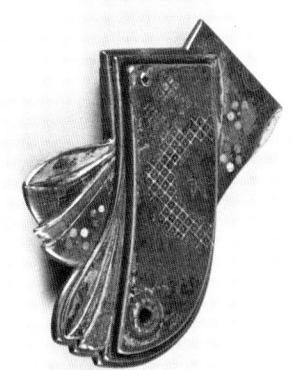

10

11 DOOR PULL
Mid-seventeenth century
Cloisonné; gilt copper alloy
2 ⁵⁄₈ x 4 ½ in. (6.7 x 11.4 cm)
Private collection, Osaka

Green and white enamels are juxtaposed to simulate the autumn leaves on a gourd vine.

12 DOOR PULL
Mid-seventeenth century
Champlevé; copper alloy
2 ⁹⁄₁₆ x 4 ¹¹⁄₁₆ in. (6.5 x 11.9 cm)
Private collection, Osaka

A platter of bamboo stalks in repoussé is the setting for a floral spray of dark blue enamels in cells cut by cold chisel work. Similar door pulls were installed in the Kuroshoin of the Nishi Honganji in 1657.

13 DOOR PULL
Mid-seventeenth century
Champlevé; gilt copper alloy
3 ¼ x 3 ¼ in. (8.2 x 8.2 cm)
Private collection, Osaka

A bronze pistil fixes two tiers of stamens and petals adorned with bright turquoise-green enamel.

14 NAIL COVER
Mid-seventeenth century
Champlevé; gilt copper alloy
2 ¹⁄₁₆ x 4 ³⁄₁₆ in. (5.3 x 10.6 cm)
Private collection, Osaka

Green enameled pine needles rise above a base of dark red enamel. A supporting sheet of copper lies beneath.

11

13

12

14

15 DOOR PULL
Mid-seventeenth century
Champlevé; gilt copper alloy
2 7/16 x 3 3/8 in. (6.2 x 8.5 cm)
Private collection, Osaka

Autumnal oak leaves are rendered in red and a variegated combination of pink, green, and white enamels.

16 HELMET-SHAPED DRAWER PULL
Mid-seventeenth century
Champlevé; gilt and silvered copper alloy
2 13/16 x 3 1/4 in. (7.1 x 8.2 cm)
Private collection, Osaka

Elegant chasing and green enamel embellish the gilt and silvered bronze helmet. A supporting gilt bronze layer lies underneath.

17 HIBACHI TONGS
Mid-seventeenth century
Cloisonné; iron, gilt copper alloy with gold, silver, and mother-of-pearl inlay
10 15/16 x 3/8 in. (27.7 x .9 cm)
Collection of Takashi Yanagi

Enameled dots of dark red, white, and blue lie within spiraling fields of green enamel. The cube-shaped head of the tongs bears heraldic designs inlaid in gold, silver, and mother-of-pearl.

15

16

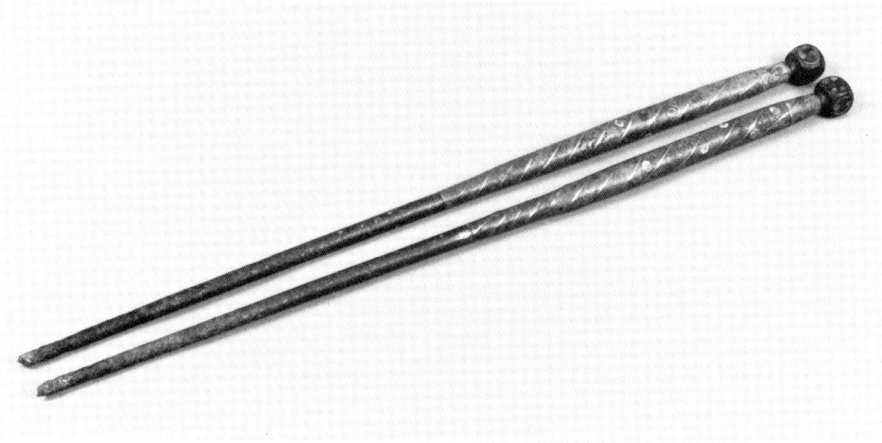

17

18 NAIL COVER
Mid-seventeenth century
Champlevé; copper alloy with gold inlay
3 1/8 x 4 in. (8 x 10.2 cm)
Private collection, Osaka

White and mottled green enamels decorate this cluster of hollyhock leaves rendered by inlaid gold wires. A whorl of chrysanthemum leaves in green enamel comprises an upper tier fixed by a central peg.

19 DOOR PULL
Second half seventeenth century
Champlevé; gilt copper alloy
3 3/16 x 3 3/16 in. (8.1 x 8.1 cm)
Private collection, Osaka

Two tiers of cherry blossoms are held in place by a gilt bronze whorl and a central peg with a loop handle. White, red, and blue enamels adorn the petals.

20 DOOR PULL
Second half seventeenth century
Champlevé; gilt copper alloy
2 5/8 x 2 15/16 in. (6.7 x 7.5 cm)
Private collection, Osaka

Green enamels in a quatrefoil and cold chiseled *nanako* adorn the recessed surface of this door pull, which is encircled by stylized chrysanthemum petals.

21 APPLIQUÉ
Second half seventeenth century
Champlevé; gilt and silvered bronze
2 1/4 x 3 3/4 in. (5.8 x 9.6 cm)
Private collection, Osaka

A sprig of paulownia is composed of dark blue and dark green enameled leaves, tiny lavender blossoms, and gilt bronze fronds.

18

20

19

21

22 APPLIQUÉ
Second half seventeenth century
Champlevé; gilt copper alloy
3 1/16 x 3 1/4 in. (7.8 x 8.3 cm)
Private collection, Osaka

This appliqué represents a paulownia spray with fronds in white enamel and leaves in green and purple.

23 NAIL COVER
Second half seventeenth century
Champlevé; gilt copper alloy
3 3/16 x 3 3/4 in. (8.1 x 9.5 cm)
Private collection, Osaka

A quatrefoil of blossoms is adorned with red and white enamels. A central peg joins the three-layered assemblage.

24 DOOR PULL
Second half seventeenth century
Champlevé; gilt copper alloy
3 x 2 9/16 in. (7.6 x 6.5 cm)
Private collection, Osaka

A floral arabesque with green enamel embellishes the outer ring of this door pull; its inner recess is incised, stippled with *nanako,* and also decorated with green enamel.

22

23

24

Sword Fittings

25 TSUBA
Signed by Hirata Hikoshirō, called Dōnin (1591–1646)
Champlevé and cloisonné; *shakudō* body, gold
2 5/8 x 2 3/16 in. (6.6 x 5.5 cm)
Tokyo National Museum

The body of this crescent-shaped *tsuba* (sword guard) is made of *shakudō,* a copper-bronze alloy mixed with gold, chemically finished to a bluish black. Various floral motifs and symbols are inlaid in gold in a design of "precious things." The lozenge-shaped blossom with red, white, and light blue enamels provides a brilliant accent to the subdued composition.

26 TWO *KOZUKA*
Hirata Hikoshirō, called Dōnin (1591–1646)
Cloisonné; *shakudō* body, gold and silver wires
9/16 x 3 13/16 in. (1.4 x 9.7 cm)
9/16 x 3 13/16 in. (1.4 x 9.7 cm)
Collection of Takeshi Wakayama

Blue, white, and red enamels and gold and silver inlays are applied to the *shakudō* ground of these knife handles to represent Mount Fuji.

27 PAIR OF *MENUKI*
Hirata Hikoshirō, called Dōnin (1591–1646)
Cloisonné; *shibuichi* body, gold and silver wires
9/16 x 1 11/16 in. (1.4 x 4.3 cm)
9/16 x 1 5/8 in. (1.4 x 4.2 cm)
Collection of Takeshi Wakayama

These *menuki,* ornaments for sword hilts, have a body of *shibuichi,* a three-to-one copper-silver alloy producing a dark gray color. Blue, white, and red cloisonné enamels decorate the metal in representations of Mount Fuji.

28 *KOZUKA*
Hirata Hikoshirō, called Dōnin (1591–1646)
Champlevé and cloisonné; *shakudō* body, gold wires
9/16 x 3 13/16 in. (1.4 x 9.7 cm)
Collection of Kunio Shimano

Gold cloisons delineate a mandarin duck and lotus richly colored with blue, red, white, green, purple, and yellow enamels.

26

25

27

28

29 SET OF SWORD FURNITURE

Mark of the Hirata school, mid-seventeenth century
Champlevé and cloisonné; *shakudō* body, gold and silver wires
Tsuba: 2 11/16 x 2 7/16 in. (6.8 x 6.2 cm)
Kozuka: 9/16 x 3 13/16 in. (1.4 x 9.7 cm)
Fuchi: 1 9/16 x 7/8 x 3/8 in. (3.9 x 2.2 x .9 cm)
Kashira: 1 5/16 x 11/16 in. (3.4 x 1.7 cm)
Tsuka: 4 7/8 in. (12.3 cm)
Kurigata: 1 5/16 x 7/16 in. (3.3 x 1.1 cm)
Origame: 1 in. (2.5 cm)
Uragawara: 1 5/16 in. (3.4 cm)
Tokyo National Museum

This set of accessories for a sword blade includes, besides the *tsuba* and *kozuka*, the *fuchi*, a sword fitting that goes next to the *tsuba*; the *kashira*, a pommel; the *tsuka*, a sword hilt; the *kurigata*, a cord knob on the side of a scabbard; and other scabbard fittings, such as the *origame* and *uragawara*. Green, blue, yellow, red, white, and mottled red and white enamels define the floral decoration.

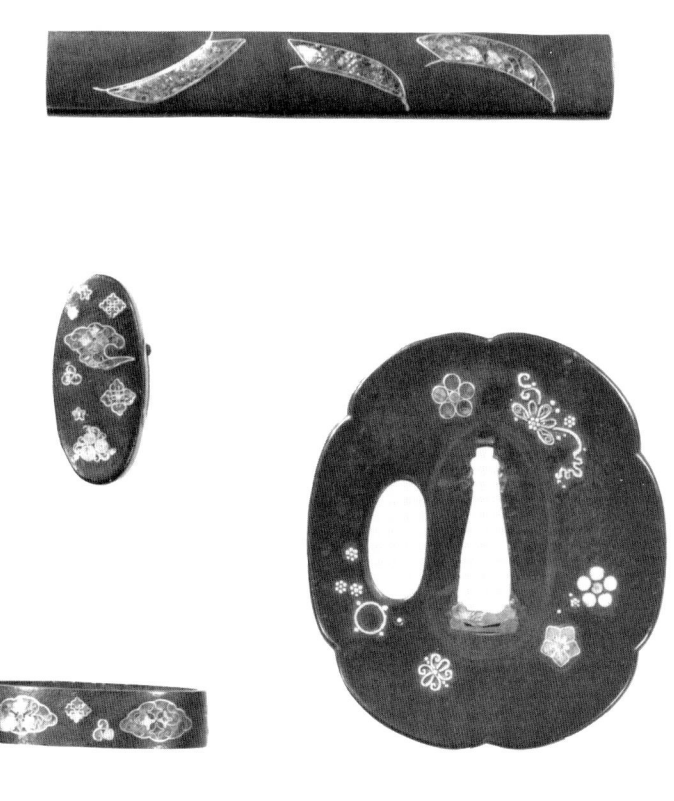
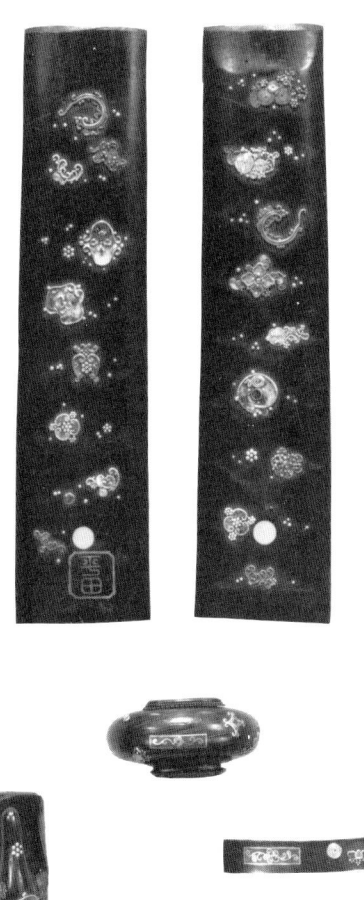

30 TSUBA
Nagasaki *shippō*, seventeenth century
Champlevé and cloisonné; *shakudō* body, brass wires
3 1/8 x 3 1/16 in. (7.9 x 7.8 cm)
Tokyo National Museum

Symbolic "treasures" are colorfully rendered with green, white, and blue enamels.

31 TSUBA
Signed by Hirata Harunari (d. 1840)
Champlevé and cloisonné; *shibuichi* and *shakudō* body, gold wires
3 1/8 x 3 in. (7.9 x 7.6 cm)
Collection of Takeshi Wakayama

Gold cloison designs filled with green, red, and white enamels adorn two facing, kidney-shaped *shakudō* fields.

32 TSUBA
Signed by Hirata Harunari (d. 1840), dated 1828
Cloisonné; *shakudō* body, silver, gold wires
2 3/4 x 2 1/2 in. (7 x 6.3 cm)
Tokyo National Museum

Dazzling crystalline snowflakes in green, white, red, purple, yellow, and light blue enamels are set against a yellow enamel surface. Along the edge of the sword guard are *shakudō* insets decorated with tiny gold cloisons and brilliant enamels.

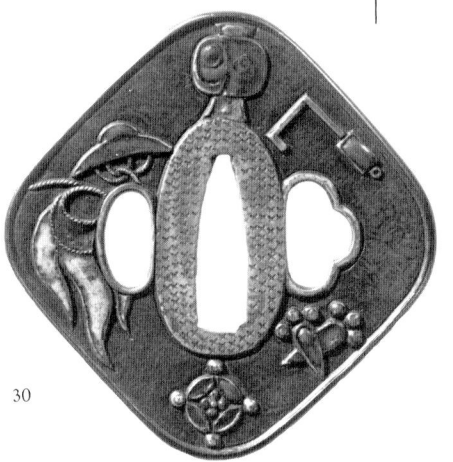

30

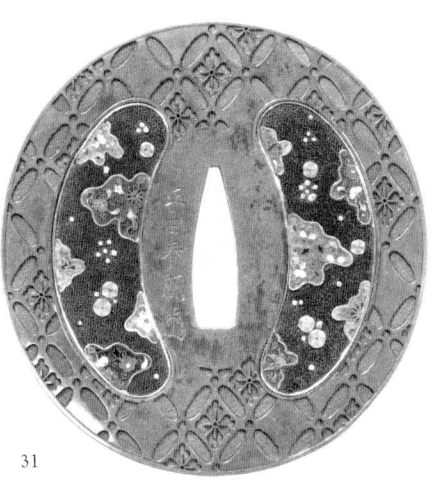

31

32

Eighteenth Century

33 DOOR PULL
Early eighteenth century
Champlevé; gilt copper alloy
2 5/8 x 3 3/8 in. (6.7 x 8.5 cm)
Private collection, Osaka

Blue, green, red, and white enamels colorfully accent this heraldic nail cover richly textured with *nanako*.

34 DOOR PULL
Eighteenth century
Cloisonné; copper alloy
2 13/16 x 2 13/16 in. (7.1 x 7.1 cm)
Private collection, Osaka

Black, blue, and red enamels fill the large circular field of this door pull. The curvilinear placement of cloisons creates a scintillating, rhythmic pattern.

35 DOOR PULL
Eighteenth century
Cloisonné; gilt copper alloy
3 7/16 x 3 7/16 in. (8.7 x 8.7 cm)
Private collection, Osaka

A Chinese-inspired blossom outlined with gilt cloisons is defined by red, white, and blue enamels against a field of dark green.

36 DOOR PULL
Late eighteenth century
Champlevé; bronze
2 5/8 x 2 11/16 in. (6.7 x 6.8 cm)
Private collection, Osaka

A central rosette of red, white, and blue is placed against a light brown enameled ground. The hexagonal recess is bordered by two intertwined bronze triangles.

33

35

34

36

Vessels

37 EWER
Eighteenth or nineteenth century
Cloisonné; copper alloy
6¾ x 11 13/16 in. (17.1 x 30 cm)
Private collection, Osaka

This round water pitcher was a challenge to the technical proficiency of its maker. It is decorated with stylized figures of a dragon and a phoenix bordered by floral motifs. The designs reflect strong Chinese influence, but the shape of the vessel is Japanese with analogies to lacquer utensils.

38 COVERED WATER VESSEL
Eighteenth or nineteenth century
Cloisonné; copper alloy
8 11/16 x 8¾ in. (22 x 22.2 cm)
Private collection, Osaka

This octagonal water container is constructed of eight flat cloisonné panels assembled into an integrated design. Dark blue, turquoise, white, and maroon enamels are employed in a design of geometrized patterns. Its interior walls and base are covered by another set of nine cloisonné panels, all of which are decorated with a central floral medallion. The joins are sealed with lacquer. The exterior bottom is covered with turquoise enamel fixed by a diaper pattern of copper wires.

37

38

Nineteenth Century

39 APPLIQUÉ
Early nineteenth century
Champlevé; gilt copper alloy
3 1/4 x 3 11/16 in. (8.2 x 9.4 cm)
Private collection, Osaka

Green, yellow, blue, and red enamels flow together freely in an interplay of color on this abstracted peony design. Fields of tiny *nanako* provide a contrasting background to the brilliant colors, and the sprightly curvilinear accents add rhythmic cohesion to the design.

40 DOOR PULL
Early nineteenth century
Cloisonné; gilt copper alloy
3 1/4 x 4 in. (8.2 x 10.2 cm)
Private collection, Osaka

Thin applications of reddish brown, yellow, and green enamels suggest a colorful autumnal display of maple leaves.

41 PAIR OF DOOR PULLS
Early nineteenth century
Champlevé and cloisonné; gilt copper alloy
2 7/16 x 3 7/8 in. (6.2 x 9.9 cm)
Private collection, Osaka

A red and yellow enameled peony blossom and a striding Chinese lion with blue, red, and white enamels are encircled by a dark green tendril.

42 DOOR PULL
Early nineteenth century
Cloisonné; gilt copper alloy
2 15/16 x 2 15/16 in. (7.5 x 7.5 cm)
Private collection, Osaka

Green, yellow, purple, and red enameled water lilies rise from a blue and white pond. The undulating cloison wires simulate the treatment of water in paintings of the Rimpa school.

43 DOOR PULL
Early nineteenth century

41

39

42

40

43

Champlevé; gilt copper alloy
4 9/16 x 3 3/4 in. (11.6 x 9.6 cm)
Private collection, Osaka

Green enamels fill a bordering row of trefoils as well as the four inner arms that connect to a red enameled center.

Modern Enamels

44 ROUND COVERED BOX
c. 1870
Cloisonné; gilt copper alloy
4 3/4 x 7 1/4 in. (12.1 x 18.4 cm)
Private collection

Geometrized textile patterns in roundels and stylized floral sprays in pink, red, white, and dark and light blue enamels are arranged on a dark green enameled ground.

45 STAND
1870s
Cloisonné; gilt copper alloy
5 1/8 x 7 7/8 in. (13 x 20 cm)
Private collection

Blue and white scale patterns and pink and red floral sprays with blue leaves are arranged against a dark green ground.

44

45

Hayashi School

46 RECTANGULAR PLAQUE
Attributed to Hayashi Kodenji II
(1859–1922), 1880s
Cloisonné; silver wires
12 x 30¼ in. (30.5 x 76.8 cm)
Private collection

A flock of white and black cranes flies through a light blue sky.

47 INCENSE BURNER
Attributed to Hayashi Kodenji (1832–1911)
Cloisonné; gold and silver wires
12⅞ x 10⁷⁄₁₆ in. (32.7 x 26.5 cm)
Private collection

Below rows of ornamental bands, cherry trees bloom on a black ground. Fine silver wires delineate the blossoms and the delicate veins of the leaves. Subtly suffused color adds tonal richness to each leaf and petal. The birds are rendered with gold cloisons.

48 VASE
Mark of Hayashi Kodenji (1832–1911)
Cloisonné; copper alloy body, silver wires
9⁷⁄₁₆ x 4¾ in. (24 x 12.1 cm)
Collection of Mrs. K. Tan

Seasonal plants that bloom at different times throughout the year are depicted together here; they include hydrangea, lilies, autumn maple, chrysanthemums, and prunus. The plants and chirping swallows are set against a polished black ground. Ornamental bands with minute cloisons embellish the neck and base.

48

46

47

Namikawa Yasuyuki

49 VASE
Cartouche mark of Namikawa Yasuyuki (1845–1927)
Cloisonné; gold and silver wires
12 x 6⅛ in. (30.5 x 15.6 cm)
Philadelphia Museum of Art

STUDY FOR A VASE
Seal of Namikawa Yasuyuki (1845–1927)
Ink and colors on paper
14¾ x 9⅜ in. (37.5 x 23.8 cm)
Philadelphia Museum of Art

The drawing bearing the seal of Yasuyuki is remarkably similar to the vase illustrating peafowl among chrysanthemums with decorative borders at top and bottom.

50 RECTANGULAR VASE
Signed by Namikawa Yasuyuki (1845–1927)
Cloisonné; silver body and wires
9¾ x 3⅝ in. (24.8 x 9.2 cm)
Private collection

Tall bamboo stalks, flowers, and a flitting butterfly adorn this four-sided black vase. The silver contour lines narrow and widen like the cursive strokes of a calligraphic brush.

50

49

51

51 HEXAGONAL VASE
Signed by Namikawa Yasuyuki (1845–1927)
Cloisonné; silver wires
12 3/16 x 5 1/4 in. (31 x 13.3 cm)
Private collection
   On a polished black vase, seasonal flowers are sensitively delineated with cursive silver lines. Subtle tonal grading enhances the rich polychrome splendor of the botanical display.

52 VASE
Signed by Namikawa Yasuyuki (1845–1927)
Cloisonné; gold wires
11 x 6 in. (27.9 x 15.2 cm)
Private collection
   Butterflies and a bird hover over blossoming dogwood, iris, and hydrangea. This coffee brown vase was made as a pendant to no. 53.

53 VASE
Mark of Namikawa Yasuyuki (1845–1927)
Cloisonné; gold wires
10 7/8 x 6 in. (27.6 x 15.2 cm)
Collection of Mr. and Mrs. Jerry Freeman
   A scene of autumn maple, flowers, and birds is composed on a coffee-colored ground.

54 COVERED VASE
Mark of Namikawa Yasuyuki (1845–1927)
Cloisonné; silver body and wires
8 1/8 x 4 3/4 in. (20.7 x 12.1 cm)
Collection of Mrs. K. Tan
   This view of the Kinkakuji in Kyoto is rendered with tapered wires and subtly graded tonal passages.

52

53

54

55 GLOBULAR JAR
Signed by Namikawa Yasuyuki (1845–1927)
Cloisonné; gold and silver wires
6 ¾ x 6 ½ in. (17.2 x 16.5 cm)
Collection of Donald Gerber

Richly colored chrysanthemums and butterflies are executed with solid or mottled enamels and set against a polished black ground.

56 COVERED VASE
Signed by Namikawa Yasuyuki (1845–1927)
Cloisonné; copper body, silver wires
9 x 4 ¾ in. (22.9 x 12.1 cm)
Collection of Dr. and Mrs. Shigeji Takeda

On this black vase, birds flutter around a garden of pink and white cherry blossoms, bellflowers, bush clover, and white chrysanthemums.

57 COVERED TRIPOD VASE
Signed by Namikawa Yasuyuki (1845–1927)
Cloisonné; silver wires
4 ½ x 4 ¾ in. (11.4 x 12.1 cm)
Private collection

The hanging wisteria on this vase is typical of Yasuyuki's later style, less naturalistic, relying more on stylization and pattern.

56

55

57

Namikawa Sōsuke

58 VASE
Mark of Namikawa Sōsuke (1847–1910)
Musen and cloisonné; silver wires
19 5/8 x 12 1/8 in. (49.9 x 30.8 cm)
Wyatt-Franke Collection

Musen is enamel from which the cloisons have been removed. It was originally developed by Namikawa Sōsuke for pictorial compositions. On this vase, doves resting on a decaying tree stump are colorfully rendered against a light brown ground. The simulated calligraphic lines are defined by bordering wires and filled with black enamel.

59 TWO-SECTION SCREEN
Namikawa Sōsuke (1847–1910), dated 1903
Musen and cloisonné; silver wires
Panels: 21 x 14 1/4 in. (53.3 x 36.2 cm)
Private collection

The lacquer screen is inlaid with four enameled panels depicting scenes entitled *Pinks in Moonlight, Pagoda in Snow, Arashiyama in Spring,* and *Wisteria over Carp*.

60 PANEL
Attributed to Namikawa Sōsuke
(1847–1910)
Musen, bokashi, and cloisonné; silver wires
15 x 19 1/4 in. (38.1 x 48.9 cm)
Collection of Dr. and Mrs. Shigeji Takeda

This wintry scene of ducks was inspired by a Watanabe painting. Green, blue, tan, black, and white enamels were used in this work.

58

60

59

Later Masters

61 HEXAGONAL VASE
c. 1900
Cloisonné; copper body, silver wires
24 ¾ x 16 ⅜ in. (62.9 x 41.6 cm)
Wyatt-Franke Collection

A hawk and doves are perched on a maple tree above flowering chrysanthemums, pinks, and peonies on this large black vase. Silver wires in various sizes are bent and tapered to simulate calligraphic lines, especially notable in the hawk. Colors and tones of enamels are manipulated to create painterly gradations. The base bears the mark of an unidentified master.

62 SQUARE VASE
c. 1900
Bokashi and cloisonné; gold and silver wires
15 ¼ x 9 ½ in. (38.7 x 24.1 cm)
Wyatt-Franke Collection

In this vase by an unknown master, four panels illustrate scenes of mandarin ducks, cranes, flowers, and jumping fish. The surrounding areas are richly decorated with ornamental brocade patterns.

61

62

63 PAIR OF SQUARE VASES
c. 1900
Cloisonné; silver wires
12 ¼ x 5 ⅜ in. (31.1 x 13.7 cm)
Wyatt-Franke Collection

Black crows delineated with cursive silver wires perch on a blossoming prunus branch. The unusual grayish green ground effectively complements the scene.

64 VASE
Early twentieth century
Musen and cloisonné; silver wires
14 ¼ x 8 ½ in. (36.2 x 21.6 cm)
Private collection

This landscape scene with a flying phoenix was inspired by Chinese paintings in the blue-green landscape style.

64

63

65 PANEL WITH A SCENE FROM TALES OF GENJI
Kawade Shibatarō (1856–1921), 1903
Musen, bokashi, and cloisonné; silver wires
24 x 24 in. (61 x 61 cm)
Private collection

A traditional Yamato-e theme, a scene from the classic *Tales of Genji,* is rendered in a rich palette of enamels. The elegant brushstrokes outlining the figures, reminiscent of the Kanō school, are executed with double bordering cloisons filled with black enamel.

66 FLOWER VASE
Mark of Kawade Shibatarō (1856–1921)
Moriage and cloisonné; silver wires
12¾ x 6⅝ in. (32.4 x 16.8 cm)
Collection of Mr. and Mrs. Jerry Freeman

Butterflies rendered in *moriage* (enamel in high relief) are arranged against a dark green ground and a field of green ferns outlined with silver wires.

67 VASE
Mark of Kumeno Teitaro (1865–1939)
Basse taille, musen, and cloisonné; silver wires
9⅜ x 4⅜ in. (23.8 x 11.1 cm)
Wyatt-Franke Collection

This scene of fish swimming under hanging wisteria is executed principally in *basse taille,* a technique in which transparent enamel is applied over relief designs in gold or silver. The artist has also made use of other techniques, utilizing each where it functioned best.

66

67

65

68 VASE

Mark of Andō Jūbei (1875–1953)
Moriage, musen, and cloisonné; silver wires
12 1/16 x 8 3/8 in. (30.7 x 21.3 cm)
Wyatt-Franke Collection

A variety of seasonal flowers are delineated with silver cloisons and rendered in moriage relief.

69 VASE

Mark of Andō Jūbei (1875–1953)
Moriage, musen, bokashi, and cloisonné; silver wires
16 3/8 x 8 3/4 in. (41.6 x 22.2 cm)
Private collection

The chrysanthemums on this vase are rendered in extremely subtle gradations of tone.

68

69

70 TRAY WITH HANDLE
Mark of Hattori Tadasaburō, active c. 1900
Moriage and cloisonné; copper wires
8½ x 10½ in. (21.6 x 26.7 cm)
Private collection

Insects and a blossoming lotus flower adorn this pale lavender-gray tray. The artist has performed the difficult feat of enameling a spherical shape without stabilizing cloisons.

71 BELL-SHAPED LAMP
Early twentieth century
Plique-à-jour
8¼ x 7¼ in. (21 x 18.4 cm)
Private collection

Plique-à-jour is a technique of enameling in which translucent enamels are held together without a supporting ground. White enamel in a tracery pattern is set against blue-gray.

70

71

Glossary

BASSE TAILLE	technique in which translucent enamel is applied over relief designs in gold or silver
BOKASHI	technique in which enamel colors are applied in subtly graduated tones
CHAMPLEVÉ	technique in which enamel is poured into grooves or areas engraved or hammered (repoussé) into metal, the fired enamels then being polished level with the surface of the metal
CLOISON	metal wire used to delineate a compartment into which enamel is poured in the cloisonné technique
CLOISONNÉ	technique in which enamel is poured into compartments formed by a network of metal wires (cloisons) attached to a metal ground
KANŌ SCHOOL	Sino-Japanese school of painting originating in the sixteenth century, the works characterized by bold angular lines and ink washes and occasionally by the addition of color
KOZUKA	small utility knife attached to a sword
MAKI-E	technique in which powdered gold is sprinkled on wet lacquer to create a design
MENUKI	ornament for a sword hilt
MORIAGE	technique in which a design is raised in high relief by the addition of layers of enamel
MUSEN	technique, used primarily for pictorial compositions, in which cloisons are removed before firing, after the enamel mixture has sufficiently stabilized
NANAKO	tiny round punch marks made with a cold chisel
PLIQUE-À-JOUR	technique in which translucent enamel is fused into a network of cloisons without a supporting ground
REPOUSSÉ	relief decoration on metal produced by hammering from the underside
RIMPA	Japanese painting style based on Yamato-e traditions, characterized by flat planes of brilliant color in decorative compositions
SHAKUDŌ	copper-bronze alloy mixed with gold, chemically finished to a bluish black
SHIBUICHI	three-to-one copper-silver alloy of a dark gray color
SHIPPŌ	enameled work of art
TSUBA	sword guard
YAMATO-E	native Japanese school of painting that developed during the Heian period (794–1185) independent of Chinese influence, the works characterized by fine lines and heavy colors